# make IT matter

## THE SURPRISING SECRET FOR LEADING DIGITAL TRANSFORMATION

PATRICIA MCMILLAN

Meaning Works

First published in 2015 by Meaning Works
PO Box 116
Spit Junction NSW 2088
Australia

ISBN:     978-0-9944935-0-7 (pbk)
          978-0-9944935-1-4 (ebook)

For information about bulk purchases please visit www.patriciamcmillan.com


Cover design by Alexander Valchev
Layout by Lu Sexton
Printed by IngramSpark

# Contents

**'The only gift is a portion of thyself.'**

Ralph Waldo Emerson

# 1 Making IT Matter

## A manifesto

I once had a dear friend who loved pretty much everything about his life except for his job. He was a charismatic, larger-than-life person to whom others were naturally drawn. He loved his family. He loved his friends. He loved his interests outside of work. He hated his job. He lived from weekend to weekend and from holiday to holiday. He was literally counting down the days until his retirement, and this was when he was in his thirties and forties, so there were still a *lot* of days in his countdown. Except that a few years ago he was diagnosed with a brain tumour, and three months later he died. I still miss him, and so do all the other people who loved him.

When I think about him I am also sad about all the time he wasted doing work he didn't love. It was a waste for him because he spent far too many hours of his life being miserable, and he missed the opportunity to do work that was challenging and satisfying. It was a waste for his employer because even though he worked hard and faithfully—he was no shirker—when a person is not engaged in their work they're not contributing their best. It was a waste for the world because he was a person of great gifts and potential, and we missed out on the amazing things he might have put into the world with that time.

I know so many people like this. People who heave a huge sigh on Monday morning as they make their weary commute to work, hanging on until Friday afternoon when they can have two days of freedom before starting the cycle again. There have been stretches in my own life when I have felt that way too.

In fact, Gallup's *State of the Global Workplace* report from 2013 shows that in Australia and New Zealand only 24 per cent of workers are engaged in their work.[1] The rest are either 'not engaged', meaning they lack motivation and are less likely to put in extra effort towards reaching organisational goals (60 per cent), or 'actively disengaged', meaning they are unhappy and unproductive at work and may spread that negativity to their co-workers (16 per cent).

I hate the idea that so many people with so much to offer, to themselves and to the world, are spending their time and their energy doing work they don't care about. It's wrong. It's an enormous waste. I am not saying that everyone should love every aspect of their work every day. There will always be parts of it we're not crazy about, and there will always be times when our motivation ebbs. But believing in your work and knowing that it matters, yes, that is possible.

At the same time, organisations of all kinds need to transform themselves to survive and thrive in a digital, disrupted world. Leading this change under any circumstances is difficult. Leading it with a team that doesn't care about making it happen is impossible. Leaders of transformation need the contributions of everyone's best ideas, unique characteristics, talents and energy.

I believe the key to leading this kind of change successfully is to bring more meaning to the work: meaning for ourselves, meaning for our teams, meaning for our customers. It's about making a real difference and knowing what that difference is. It's about believing in what you do. It's about making it matter.

This means we need to change our conversation from one that speaks only in data and analysis, technology and processes, to

1 The percentage of engaged workers is higher in Australia and New Zealand than in most other places. Worldwide, it's only 13 per cent.

one that speaks to people at a deeper level—to their experiences, feelings and aspirations. Paradoxically, as the world becomes more digital and automated, we create the greatest value by being more human. It's time to get personal about work.

## The dragonfly

Why is there a dragonfly on the cover of a book about the rather odd combination of digital transformation, information technology and storytelling? Let me explain.

The dragonfly is a symbol of change and transformation because it begins its life in water and then moves into the air and flies. Digital disruption calls for your organisation to change and transform, and building the skills you need to lead this change may call for you and your team to transform.

The dragonfly is a symbol of agility. Dragonflies can change direction very rapidly, with no apparent effort, just as agile organisations and teams do.

Because they are found near water, a symbol of emotional depth, dragonflies are associated with the idea of looking beneath the surface for deeper meaning. As we'll explore in this book, stories are a tool for looking beneath the surface to reach a greater depth of understanding.

Finally, because of their sprightliness and translucence, dragonflies are a symbol of lightness and joy. Transformation, change and meaning might sound like heavy topics, but they don't need to be. In fact, change is much more effective when it's approached with lightness, positivity and humour.

Transformation, change, agility, emotional depth, meaning, lightness and joy. Sounds to me like the perfect symbol for the ideas in this book. May the spirit of the dragonfly be with you.

## Who this book is for

I wrote this book with an information technology audience in mind, especially chief information officers (CIOs), digital strategists and their teams. There are a couple of reasons for this.

One is that I worked in IT for many years, first as a programmer and later as a business analyst, a project manager, a team leader and a director of strategic initiatives. For much of that time I worked within organisations where IT was a support or enabling function, not the core business, and this means I have firsthand experience of the fractious relationship IT usually has with its internal customers. I have a great deal of empathy for hardworking IT teams who are blamed for every problem, even when it's not their fault, and who receive no recognition when things run smoothly because no one pays attention when everything is working as it should. I also have enormous respect for the ingenuity, dedication and just plain delightfulness of many IT professionals I've been privileged to work with over the years. The world needs you.

The other reason this book was written with an IT audience in mind is that I believe CIOs and their teams are facing particular opportunities and challenges related to digital transformation.

It wasn't too long ago that the CIO's job in an organisation was to keep things running as smoothly and efficiently as possible. IT was a support role, not a lead player and usually not a strategic partner. However, as organisations recognise the need to reinvent themselves for a digital, automated, customer-driven, hyper-connected world, this is changing. Technology is suddenly at the forefront of organisational strategy.

As a consequence, CIOs are being asked to do much more than support operations; they're being asked to enable and sometimes lead digital transformation, and often the survival of the organisation will depend on how well this is done. Not that there's any pressure.

Like all great opportunities, this one comes with some big challenges, and the technology is the least of them. If you were to ask people in your organisation whom they would trust to lead a major transformation to help the business thrive in the future, most of them would not point to the IT department. If you were to ask them who has the vision and the communication skills to inspire meaningful change, most would not name the CIO. If you asked them who is the greatest advocate for the needs of the customer and has their finger on the pulse of what the customer wants, the IT team would not leap to their minds.

In other words, CIOs' biggest challenges relate to communication, relationships, vision, culture and, well . . . people.

This book will give CIOs and their teams a new perspective on how to approach these challenges, and some tools that can make a real difference.

If you don't work in IT but you are facing or leading any kind of change or transformation in your organisation (and who isn't?), this book is for you too. Although I refer to IT from time to time, the same principles apply to any kind of team or function in just about any organisation. So don't be put off by the IT context. The need to make work meaningful—to make it matter—is felt by everyone, and that's what the tools in this book are about.

## What is digital transformation?

We're living in a time of radical change. There's no stopping the clock, and no going back. All types of organisations are undertaking digital transformation programs to ensure their continued survival.

A definition is a useful place to start, so here is one from the Wikipedia article on digital transformation:

> Digital transformation refers to the changes associated with the application of digital technology in all aspects of human society. Digital transformation may be thought as the third stage of embracing digital technologies: digital competence → digital literacy → digital transformation. The latter stage means that digital usages inherently enable new types of innovation and creativity in a particular domain, rather than simply enhance and support the traditional methods.

A few drivers in particular are contributing to the recent scrambling of organisations to reinvent and reimagine themselves for a digital world.

Digital disruption is forecast to displace nearly half of all Australian jobs in the next 20 years. According to the 2013 Deloitte Access Economics Report *Digital Disruption: Short fuse, big bang?*, the industries expected to face both significant and imminent disruption by digital include finance, retail trade, arts and recreation, professional services, real estate, and information, media and telecommunications. By 2018, add education, agriculture, government services, health, transport, recruitment and utilities to this list.

With brokering services available for almost everything from hotels to higher education, this has become the Age of the Customer. Customers have more choice now than ever before, and they have access to a global online marketplace. Organisations that are not providing the right customer experience are not going to survive.

Smart machines are doing all kinds of work that used to be done by humans. It may not be long before your doctor is a robot.

The Internet of Things means just about everything will soon be able to talk with just about everything else through embedded electronics, software, sensors and network connectivity.

These and other trends mean that suddenly technology is at the forefront of organisational strategy—and IT, which not long ago may have been a support or enabling function, is front and centre. Your organisation needs you to help lead them through digital transformation, and this means leading big changes across the business. The CIO is in the hot seat. It's exciting, but scary.

## Why leading digital transformation is hard

Leading any kind of change, anytime, anywhere, is hard, but CIOs are often starting on the back foot.

Fairly or unfairly, perceptions about the IT group can be pretty grim. According to the results of *CIO* magazine's State of the CIO 2015 survey, 53 per cent of CIOs believe that IT is seen by the wider business as an obstacle to their mission. And these are just the ones owning up to it.

This conclusion is echoed by comments like these that I have heard from people about their organisation's IT group:

'IT is in the basement. Literally in the basement. You only hear from them when things go wrong. You get these emails saying the network is about to go down . . . again. Or a system is going down . . . again. And I only go to see them when things go wrong. I go down to the basement and tell them, my card's not working . . . again.'

'IT is a black box. We sort of throw things over the fence to IT and then we don't know what's happening with them. A couple of months later, maybe something comes back.'

'They speak in gobbledegook and I have no idea what they're talking about. Can they please just tell me what it means for me?'

In an August 2015 report from Forrester Research, analyst Bobby Cameron identifies 'CIOs' Top Five Career-Ending Digital Transformation Challenges'. To thrive in digital transformation, Cameron says, CIOs need to be able to drive big-picture strategy.

He says the main obstacles are:

1. You're not trusted.

2. You can't inspire meaningful change.

3. You're stuck in technical debtor's prison.

4. You only think small.

5. You aren't customer-obsessed.

Notice that only one of these challenges relates specifically to technology. The rest are about human-centred skills: communication, engagement, vision, relationships. The key word in digital transformation isn't *digital*. It's *transformation*. Fundamentally it's not about the technology. It's about people and changing behaviour, and if CIOs are to be partners in leading the change, they need to start by transforming the way they engage and communicate.

## Crossing the gap

Transformation means change, and change means changing people's behaviour.

One way to try to change behaviour is to push things on people and try to force them to change.

This has been IT's usual tactic in the past. 'Trust us, we're from IT. We know technology better than you. Now just use the new system. It will be wonderful, and it will solve all your problems.' But it turns out the new system doesn't match what the customer needs, it breaks other things that were important, and it generally makes everyone's life worse.

This approach of pushing things on people doesn't win you any friends and is one reason CIOs haven't earned trust.

Another way to change behaviour is to try to get people to work with you to bring about the changes themselves. A pull strategy rather than a push strategy. This is much better, don't you think?

To make this work there needs to be a story people tell themselves, a story they want to be part of. It needs to mean enough to them that they're willing to make changes, changes that may be difficult, challenging, downright frightening.

Facts and analysis don't provide this story. It has to be more personal than that. This story has to speak to people's feelings. In his book *The Heart of Change*, John Kotter, one of the world's leading authorities on change, says, 'People change what they do less because they are given *analysis* that shifts their *thinking* than because they are shown a *truth* that influences their *feelings*'.

How do you find these truths? That's the subject of the rest of this book.

First, there are four tasks you need to undertake.

# 2 Four Tasks

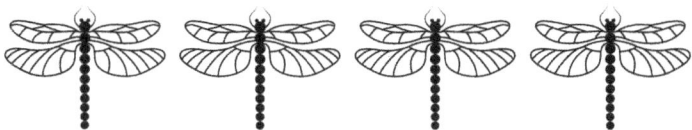

## Four tasks

In traditional stories, the hero must complete a series of difficult tasks in order to achieve the reward—to save someone they love, to get home, to vanquish evil. You are the hero of this book, and your story is no different. (We'll talk more about the hero's journey in chapter 7.) To save your organisation and help it to thrive through a transformation, there are four tasks you must undertake.

The first three tasks are questions you need to explore. You need to find out why what you are doing *matters*. Find out why it matters to you personally. Find out why it matters to your customers. Find out why it matters to the people who will make it happen. Without answers to these questions, no one has any reason to change what they do, which means there's no change and no transformation. Knowing why it matters isn't enough, though. You also need to put it into action, and that is the fourth task.

These tasks don't need to be done in sequence. In fact, you'll find that answering any one of these three questions will help you answer the others, and that the actions you take, if you reflect on them, will also help you find the answers to the questions.

## Find out why it matters to you

Your first task is to find out why it matters to you.

Have you seen Simon Sinek's TEDx talk 'How great leaders inspire action?' If you haven't, or if it's been a while since you have, do yourself a favour and watch it. It's one of the most popular talks on TED, with over 24 million views at this writing. It's so popular because Sinek has a very simple but very compelling message: start with why. People don't buy *what* you do, he says, they buy *why* you do it.

He uses Apple as an example. He asks why Apple can be so innovative year after year when they're just a computer company like any other. Keep in mind the talk was filmed in 2009.

All great leaders and companies communicate the same way, he says, and it's the opposite to everyone else. Everyone knows what they do. Many know how they do it. But very few know why they do what they do, and the why is what makes all the difference.

'If Apple were like everyone else,' Sinek says, 'a marketing message from them might sound like this: *We make great computers. They're beautifully designed, simple to use, and user-friendly. Want to buy one?* Meh. But here's how Apple actually communicates. *Everything we do, we believe in challenging the status quo. We believe in thinking differently. The way we challenge the status quo is by making our products beautifully designed, simple to use, and user-friendly. We just happen to make great computers. Want to buy one?*'

The difference is that Apple communicates what they believe and stand for, and they attract customers who believe the same things.

This principle is not just true for companies. It's true for individual leaders as well. It's true for you. If you know why you do what you do, why it matters to you personally, and you communicate that to others around you, this changes the whole conversation. Find and communicate your 'why', make sure it's real and make sure your actions match it, and you will earn the trust and support of your team, and the trust and support of others in your organisation who believe the same things you do. And you'll find there are more of them than you might have thought.

The other thing this does for you is to bring clarity and meaning to your own work. Once you know why your work matters to you, you'll find it much easier to identify the things that are important, the things that need your focus and attention, and you'll find it much easier to let go of the things that don't matter so much. It makes your work simpler and at the same time more satisfying.

There are two reasons this task is difficult. In fact, it may be the most difficult of the four tasks you face. The first is that you have to do some searching through your experiences and your beliefs to find out why your work matters to you. This is an ongoing process, and the first few times you try to express it, it might sound thin and unconvincing. Keep going until you've found something that rings true. It can take a while. You'll find some help on this in chapter 5.

The second reason it's difficult is that once you have figured out why your work matters to you, you need to communicate those beliefs to other people. You can't just keep them hidden away for yourself. Others need them too, and saying them out loud to other people will also help you find out what's true. This requires you to reveal something real about yourself to the people around you at work—your team and your peers. It requires you to be human and open, and this can be very uncomfortable until you practise it and see the amazing results it produces. If you're having trouble coming out of your shell, check out Brené Brown's TED talk, 'The Power of Vulnerability'.

## Find out why it matters to your customers

Remember Bobby Cameron's list of CIOs' top five career-ending digital transformation challenges? Number five on his list is 'You're not customer-obsessed'.

In the Age of the Customer, people have more choices than ever before about what they'll buy, who they'll buy it from and how they'll buy it. This is one of the major disruptors organisations are facing. Hotels are being disrupted by Airbnb, taxis are being disrupted by Uber, universities are being disrupted by online learning providers. People have more choice.

Even if your customers are internal and you think they don't have a choice about whether or not they use your services, think again. If you are seen as an obstacle, if your service is perceived as too slow, your processes as too complicated, or your outcomes as not adding value, your customers can usually find a way around you to get what they need. The IT groups I work with like to refer to this as 'Shadow IT'—the IT services their internal customers procure without consulting central IT. You can bet your customers don't call it Shadow IT. They just call it an easier way to get things done.

The way to fix this is not to introduce stricter procurement policies, but to understand your customers better so it's easier for you to provide products and services that make a real difference to them. Without that understanding you're just guessing.

Just as you have your own 'why' about your work, your customers have theirs. They do what they do for a reason; there are things that are important to them. Your task is to get to know them well enough to find out what these are.

Too often when we talk with customers we're only asking about their project requirements or sending them a customer satisfaction survey. It's rarer to take the time to understand our customers' broader context—what they do, how they go about doing it and, most importantly, why they do it.

I did some work with a group of software engineers who were building an app for a medical researcher. They'd been working with her for months, yet when I asked them what her research was about, they didn't know. They knew what features the app was meant to have and how it was meant to work. They knew she had won awards for her research, but they didn't know what the aim of her research was. They didn't know her 'why'. When I went to speak with her and came back to tell them about it, they thought it was very cool. I bet if they had known this earlier, it would have saved them a few iterations of development and she could have had a product that delighted her much sooner.

Airbnb knows that for their customers, travelling isn't just about having a nice hotel room at a reasonable price. It's about exploring the world, meeting real people and having a sense of belonging, no matter where you are.

Finding these things out will help you provide your customers with something of real value to them, something they will love, something that makes a difference. Chapter 6 will help you with this task.

## Find out why it matters to the people who will make it happen

If you've explored the first two questions you have an idea of what you believe and what's important to your customers. Your next task (but remember, they don't need to be done in order) is to find out why it matters to the people who will make the transformation happen—your team, your peers, the people who control the budget, and the decision-makers and influencers in your organisation.

This task is about helping the people around you to find their own meaning in the transformation. Everyone wants to know that the work they are doing is important, that it makes a difference. This is something people hunger for, and helping them to find this perspective, helping them to see how their work fits into a bigger story, is one of the most important tasks you can do as a leader. This is what it is to inspire meaningful change.

Too often, strategies are communicated in terms of their benefits for the organisation, but these may not matter to the people who will make the change happen. For example, the strategic plan might describe a revenue goal (to become a $500 million company) or a goal of becoming one of the top five organisations of its kind in Australia. Why would the people working there be motivated by that? The executives may find that compelling because being part of making it happen will look good on their CVs and advance their careers. But it doesn't mean anything at all to the people on the ground.

With a digital transformation, particularly one that includes more automation, there's something else at stake for the people who will need to make it happen: their jobs. The nature of the work available in the organisation is likely to change, possibly significantly. Some jobs may become redundant. Other opportunities will open up. It's critical that you find a story they can be excited about, not one that sounds like it's good for the organisation but puts them out of work.

If you want them to be inspired by the idea of the transformation, you have to put it in terms that are meaningful to them as individuals. You have to know what kind of story they want to be part of.

The answers you've discovered about why your work matters to you and what's important to your customers will provide big clues to a larger story that everyone will find compelling. But people have different roles to play in this story. Think of *The Lord of the Rings*. The larger story is the struggle to defeat the evil sorcerer Sauron by destroying the One Ring so Middle-earth can be saved. Every character fits into this larger story, but each of them plays a different role, from Frodo the Ring-bearer, to Gandalf the wizard, to Sam Gamgee the faithful friend, to Aragorn the wanderer, to Gimli the craftsman. They all contribute their unique talents and characteristics based on who they are.

In this task you need to shape the bigger story of the change, then allow the others around you to find their own role within it. Everyone wants to be the hero of their own lives. You're helping them to find out what kind of hero they're going to be.

This requires listening and empathy, to help you pick up the clues to what's important to each of the people around you. It also means beginning to think and talk in terms of this larger story, which is fascinating even though it may feel uncomfortable at first. An understanding of how to shape the larger story will help, and you'll find more on this in chapter 7.

## Put it into action

Okay, you've been soul-searching and reflecting and listening and empathising and talking as you sort through the questions of why it matters to you, why it matters to your customers and why it matters to the people who will make it happen. You may be asking, when are we going to do something?

Now. Now is when you put it into action. Make a plan and do the first sprint.

I want to emphasise that you shouldn't wait until you've completed the other three tasks before you act. Because if you do, you'll never start. Don't wait. Act. Right now is the perfect time. But as you act, keep those three questions in mind and keep working on them.

Clarity and motivation both come through action. Doing something gives you and your team a sense of progress, which shows them they *can* do something, which motivates them to do more (especially if it worked). Doing something gives you a way to test out your current thinking about what matters and how to communicate it. This gives you more clarity to refine your thinking.

This is a lot like using an Agile methodology with a project. You do a sprint, you test it out with people, you learn something that helps you do the next sprint a little better. By the end of the project, you and your customer finally know what the requirements *really* were, when neither of you would have been able to articulate them very well at the beginning.

It's the same with creating meaning. Your first version of the big picture, the vision, may not sound quite right. But as you continually test it out through acting on it, you learn more and more about it until you have one that resonates with you and with everyone around you. How many years did it take Apple to know that their vision was 'Think Different'? Great ideas and great stories may look like they have sprung fully formed out of nowhere, but they haven't. They take trial and error, and reflection.

Your task for making it matter as you put your plan into action is to weave the stories of your team's successes and failures into the larger story of the transformation. Don't lose these stories and don't hide them. They are a big part of letting your team experience meaningful progress, and of letting the rest of the organisation see the impact of the change. They are also how you and your team can learn from your experiences. Chapter 8 will help you with this task.

# 3  The Enemy

## The enemy

In any good story, the hero faces fearsome enemies—a dragon, an evil sorcerer, a soulless corporation, a vicious storm. Often the enemies that are most difficult to face are internal: the hero's own fears and beliefs.

Despite what you may think, you main adversary is not the Chief Marketing Officer. The CMO is your friend. Really.

One major adversary you will encounter in making work more meaningful is the idea of speaking a more human, emotional language rather than one rooted in data and analysis. This shift can be very uncomfortable because it requires more openness and vulnerability than you may be accustomed to displaying at work. This is the adversary inside yourself you need to confront.

There is another enemy you'll face as you try to build trust and inspire meaningful change: shadow stories.

## Shadow stories

Shadow stories are the stories people tell themselves and other people that run counter to your messages of transformation. They're a bit like cockroaches in that most of them are hidden. For every one in the open you can bet there are a hundred others living in the walls, breeding like crazy. Australia and New Zealand have relatively polite business cultures, and only the most straightforward people will tell you directly what they are thinking if it's unpleasant for you to hear.

Some shadow stories people might be telling themselves are these:

'He's been promising the same thing for two years and nothing has happened. There's no reason to believe him now.'

'The IT group don't have any understanding of what we really do. They spend a lot of money running big projects that don't bring us any benefit.'

'They're just using us until the new automation systems are up and running. Then we'll all lose our jobs.'

'They tell us self-service is about working smarter not harder, but it just seems to mean I don't have my admin staff anymore and I have to work longer hours to get things done.'

'I'll nod as if I'm going along with it, but really I think it's a bad idea. I'm not going to take any action on it.'

'There's just something about her I don't like.'

'Their new tagline is Exceeding Your Expectations. What a joke.'

Shadow stories express things that people are mistrustful, fearful, angry, frustrated or confused about. Shadow stories stop people from making positive changes. They can bring any transformation, project or good idea to a screeching halt.

Being aware of the things that attract shadow stories can help you to avoid them, neutralise them, seed more positive stories to replace them, or even just bring them into the light so you can talk about them openly.

## When your actions don't match

Shadow stories form when your actions don't match your words. Take a self-inventory to make sure you are walking your talk. Sometimes it can be hard to spot our own hypocrisy, though, so ask some trusted colleagues to point it out to you. Nuff said.

## When your stories are only partly true

This one is similar. Shadow stories form when the stories you tell are not authentically true. These are success stories that are a little too shiny. They may be based on fact but they give the wrong impression about what really happened.

Here's a famous example: 'Seven hundred happy passengers reached New York after the *Titanic*'s maiden voyage.' The facts are true. More than seven hundred of the *Titanic*'s passengers did reach New York, and they were almost certainly extremely happy to do so. However, the story fails to mention that the *Titanic* sank and fifteen hundred passengers drowned. When these facts become known it will generate a huge backlash, and plenty of shadow stories.

Don't mislead people.

## When you don't connect

Here is one that may surprise you but is extremely important to know about. Shadow stories form when you present analysis or information to people, but you haven't first made an emotional connection with them. Understanding this can be a game changer for highly analytical people everywhere—like many people who work in IT.

We like to think the facts speak for themselves, that we are rational creatures who make decisions based on logic and reasoning. In fact this is not what happens. We make decisions based on our feelings and intuitions, and then we find very logical reasons (and we can always find some) that support the decision we have made on instinct. This means if you warm people up by making a human connection, you are much more likely to earn their support. If you launch straight into cold analysis, you are likely to generate resistance.

In *Thinking, Fast and Slow*, Nobel Prize winner Daniel Kahneman explains this phenomenon. Psychologists have identified two modes of thinking, System 1 and System 2, which Kahneman portrays as characters in a story. System 1 is fast and intuitive, and we have no control over it. It uses our prior experience and memories, our feelings and associations, to reach very rapid answers. It is a system for jumping to conclusions, for creating stories and shadow stories. System 1 works all the time, automatically.

System 2 is slow and deliberate. It reasons and makes conscious decisions. It's what we like to think of as 'us'. But it works only when we force it to. Kahneman calls it 'the lazy controller'. When you are thinking through a problem logically, System 2 is at work.

System 1 is a marvellous thing. It puts experiences and associations together and comes up with ideas, and to us it feels as if they have come from thin air. We would have no creativity, no innovation, without System 1. It's also essential to our survival because it lets us get through the day without having to think carefully about everything we do.

System 1 is constantly suggesting ideas to System 2. Most of the time, System 2 just accepts these ideas because it's too much effort to test every conclusion to make sure it's correct. The easier thing for System 2 to do is find reasons to support the suggestion System 1 has made. Although System 2 is the controller that consciously makes the decisions, System 1 is highly influential—more influential than we are usually willing to admit.

Suppose you are presenting your business case in a meeting. You present the problem, some analysis, some possible solutions and your reasons for selecting the solution you are recommending. Meanwhile, there's something about you that reminds one of your audience members of their childhood dentist—not a happy memory; another has negative associations with the colour of your shirt; and another feels the tone of your voice is argumentative. Their System 1 creates a shadow story, telling them they don't like your idea, and their System 2 kicks in to find all the perfectly logical reasons why your idea won't work. Suddenly what you thought was going to be a straightforward meeting turns into a skirmish.

What you need to do instead is to get on the good side of their System 1 by taking the time to make a connection with them. Be open and tell them a story about yourself to build trust. Be self-deprecating and make them laugh. Find some common ground. Ask them about themselves (and be genuinely interested in their responses). If you do this, their System 1 will like you and will speak to their System 2 on your behalf, and you'll stop some shadow stories from forming.

## When people don't understand you

Shadow stories form when people don't understand what you're saying. If you make them work too hard to figure out what your message is they'll form their own story, and it may not be one you like. This happens when you speak in jargon, acronyms, tech speak and meaningless business abstractions.

Again, it can be hard to recognise you are doing this. Everyone in the world is subject to a cognitive bias known as the curse of knowledge. The curse of knowledge is that once you know a thing, you can't un-know it, and this prevents you from viewing information from the perspective of someone who doesn't have the same knowledge. This means sometimes you fail to make sense to those who are less informed, which is why a newbie or outsider can often give a better explanation of a topic than an expert. They can still remember what people don't know.

As an example, here's a fairly typical mission statement. This one happens to come from Stephen Covey's book *The 7 Habits of Highly Effective People*. Let me know when you've figured out what it means.

> Our mission is to empower people and organisations to significantly increase their performance capability in order to achieve worthwhile purposes through understanding and living principle-centered leadership.

Why do we speak in meaningless gobbledegook so frequently?

It's not because we intend to confuse people, but because we're trying to distil complex ideas into the fewest number of words. So we use abstract language as a kind of shorthand. Instead of simplifying, however, abstraction just makes it hopelessly confusing for anyone who doesn't have the same context.

This is what your customers mean when they say you speak in tech speak, rather than in language they understand.

To make sense and to be memorable, messages need to go the other way. They need to be more concrete. For example, do you remember the description Apple used when it introduced the iPod? Was it 'portable digital music player with a 5 GB hard drive'? No. It was '1,000 songs in your pocket'. Concrete.

It can be difficult to know when you are slipping into abstract or tech speak, because what you are saying makes sense to you; it just doesn't make sense to anyone else. When you see confused looks on other people's faces you can bet this is what's happening.

Make the effort to make your messages clear and engaging so people don't have to work to understand them. Use more stories, real examples, metaphors, diagrams and pictures to convey what you really mean and avoid some shadow stories.

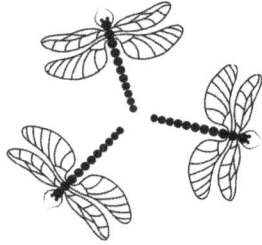

# 4  A Magic Object

## A magic object

The hero in a traditional story is often given a magical gift for their journey. The gift will somehow help with the tasks that must be completed or with the adversaries that must be overcome, although the way it will help isn't clear at the beginning.

Your gift for this journey, your magic object, is story. Your own stories, your team's stories, your customers' stories. Stories you'll listen to and stories you'll tell. It may not be clear at the beginning how these will help, but trust me, they will. Imagine you're Jack and you've just sold your cow to a mysterious old man in exchange for some magic beans. What these magic beans will do for you may be unclear at the outset, but it turns out there's a lot that happens when you plant them. Which is an apt way to introduce the metaphor of stories as seeds.

# Stories are like seeds

I am very fortunate to live near Balmoral Beach in Sydney, which I think must be one of the most beautiful places in the world. The promenade along the beach is lined with huge fig trees. Every year when the trees fruit, the birds and fruit bats have a party, gorging themselves, and of course little figs end up splattered all over the footpath and get stuck on the soles of people's shoes.

The seeds in a fig are tiny and very easy to move from one place to another. Each one of them carries the whole pattern of that massive tree, and if you plant one in the right conditions it will grow into a fig tree very much like the one it came from.

Stories are like these seeds.[2] Every story carries the whole pattern of a person's experience—an event in that person's life. Like a seed, a story is a small thing that's easy to move around and to store. It's easy to remember, and one person can tell it to another, who can tell it to another. When it's told, it unpacks the whole experience and brings with it a huge amount of meaning and context. It's as if the listeners have had that experience themselves.

A number of studies have shown that this is actually what happens in our brains when we watch or listen to a story. For example, in one experiment Princeton neuroscientists Greg Stephens, Lauren Silbert and Uri Hasson[3] used functional magnetic resonance imaging (fMRI) to record the brain activity of a speaker telling an unrehearsed, real-life story and the brain activity of a listener

2 Thanks to Cynthia Kurtz for this metaphor from her book *Working with Stories in Your Community or Organization*, Kurtz-Fernhout, 2014.

3 Greg Stephens et al. (2010). 'Speaker–listener neural coupling underlies successful communication'. *Proceedings of the National Academy of Sciences of the United States of America*, vol. 107, no. 32.

listening to a recording of the story. They found that the listener's brain activity mirrored the speaker's brain activity, with a short time delay. Other studies have recorded brain activity while volunteers are watching films.

Listening to a story is not passive; it's as if we're participating in the experience ourselves. And the more compelling and detailed the story, the greater our level of participation in it. In this way we can incorporate others' experiences vicariously and learn from them almost as much as we learn from our own direct experience of the world, an ability that may have helped humans survive over the millennia.

This is why stories are such excellent vehicles of understanding between people. When you tell a story you are not just providing people with information (information that's hard to make sense of and easy to forget). You're providing them with a seed that contains a whole experience, a memorable experience where insights and understanding come along for the ride.

## The story continuum

What exactly is a story? When story nerds like me try to nail this down the answer can become unnecessarily complicated, but you've been surrounded by stories all your life. You have an intuitive sense for what feels like a story and what doesn't.

Pretty much everyone agrees that a story involves a sequence of events (real or imagined) that happened (or will happen) to someone (real or imagined) in a particular time and place (real or imagined). So a story is something that happened to somebody, somewhere, sometime. Then the disagreements start to arise— about the elements a story needs to contain and how it should be structured, or whether the definition should focus instead on the effect stories have on listeners or on how stories are conveyed and spread. This is too much detail for our purposes, although I find the subject fascinating and I'm always happy to explore it over a couple of drinks.

Here's what I think will be helpful for you. I like to think of stories on a continuum. The stories on the left of the continuum are naturally occurring stories. These are stories you tell to the people around you every day without thinking about it: what happened at school, what happened at work, what happened on your way to meet your sister. You're not thinking about how to structure these stories; you're just thinking about how to communicate your experiences to other people, and stories are the natural way to do this. Switching metaphors slightly from seeds to plants, these stories are like plants growing wild in their natural habitat.

All the way over to the right of the scale are highly sculpted stories: novels, movies, *Game of Thrones*. These are stories that are carefully constructed in order to get the maximum engagement and response from people. They have all the right elements and the right structure to keep you turning the pages or glued to the screen. These stories are like a carefully planned and manicured garden. The best ones are like a beautiful Japanese garden, where everything feels completely natural even though every detail has been placed there with care and purpose.

The stories we're most interested in are somewhere in the middle. When you are listening to other people's stories—your team's or your customers' stories—these will be very close to stories in their natural habitat. They're just a step to the right because you're prompting for particular types of stories, so there's a shared awareness that this is what is happening. They're kind of like plants or seeds you've found in the wild and planted in your garden, but without doing any pruning or shaping.

When you are telling stories to other people for a purpose, you'll be shaping them a little, to make sure they're engaging, to convey a particular meaning, to inspire action or to spark ideas. We'll talk more about how to do this in chapters 7 and 8. These stories are a little further to the right on the continuum, but they still need to be authentic. They are like plants in your garden that you've pruned a little to give them a pleasing shape. But they're not topiary.

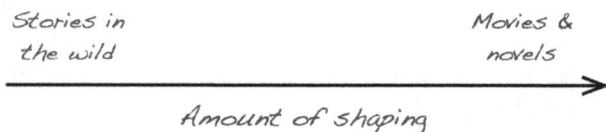

Stories in
the wild

Movies &
novels

$\longrightarrow$

Amount of shaping

## What stories do

Here are some scenarios to give you a feel for how this story thing works, and how stories can help you to build trust, engage with your customers and inspire meaningful change.

Let's say someone listens to a story you tell about why the work you are doing is important to you. (For example, have a look at the story I told at the very beginning of this book about my friend who hated his job.) As they hear this story, they identify with it. They may think of similar experiences they've had in their own life or times when they have felt like this too. They see you're human, that you have things in common with them. They like and trust you a bit more than they did before. They remember this story more than anything else you said to them. (Now don't spoil this good will by acting in a way that's not consistent with your story.)

Okay, another scenario. Let's say your team listens to a customer's story. As they hear it, they identify with it. They start to empathise with the customer and understand their perspective more clearly. They start thinking of new ideas for how your team's services can be tuned to meet the customer's needs more effectively, or new ideas for products the customer would love, or ideas for ways you can collaborate. You've created a much more effective customer focus and you've stimulated innovation.

Now let's say you tell your team a story about a real person who has actually succeeded with some aspect of digital transformation. Your team identifies with the person in this story. They like how the story turned out for that person. They learn from what that person did, and it gives them ideas and confidence about what they might be able to do for your organisation's digital transformation. They start to put their new story into action. You've inspired meaningful change.

Do you see where we're going with this? The stories generate insights, which can turn into ideas, which can turn into actions (and the actions generate new stories). It's simple, but it's amazing.

## Four stories

There are four categories of stories that will be useful to you in making work more meaningful and in successfully leading digital transformation: identity stories, insight stories, influence stories and impact stories.

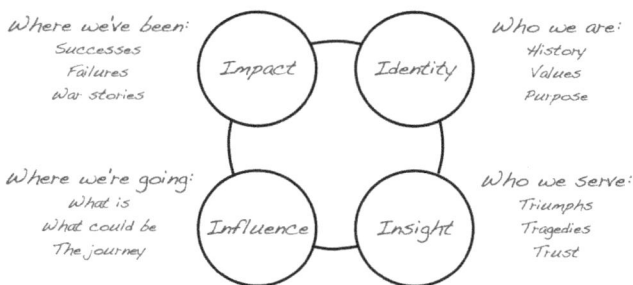

Identity stories are about who you are. This can be who you are as an individual, to help you as you tackle the first task of why it matters to you, or it can be who you are as a team or an organisation. Identity stories include stories about your history, your values and your purpose.

Insight stories are about who you serve—your customers. They're about who your customers are, what their purpose is, what they love, what they find rewarding, what frustrates them and drives them crazy, and what their past experiences have been with your organisation or your services.

Influence stories are the stories that inspire change. They tell the story of where you are going, the journey you are embarking on—

how it will be along the way and what it will be like when you get there.

Finally, impact stories are about where you've been. Again, these can be about you as an individual, or they can be about the team. Impact stories demonstrate the successes you have had. They are your case studies, the stories you want to share with everyone who will listen. They are also your failures and war stories, which make sure important information about what you've learned isn't lost.

In the rest of the book we'll explore how to discover and draw out these stories, how to shape them and what to do with them once you have them.

## Finding stories

There are a number of ways you can find stories from your own experience, from your team and from your customers. The sections that follow contain prompts and questions to help you draw out particular types of stories. Here are a few methods you can use with the prompts; you can probably think of others.

### 1. You can recall your own stories from your life experiences.

A note about digging for your personal stories: don't expect them just to appear on command. They can be a bit skittish. They will occur to you unexpectedly at odd moments; when they do, note them down somewhere so they don't slip away again. I use Evernote to keep track of things, but it really doesn't matter what tool you use. You don't need to write out the story. All you need is a title and a couple of bullet points to help you recall what happened. For example, you might have a story called 'The time the flood hit our town', with a note or two about the significance of the story for you or how you think you might use it. A good way to entice your stories from their hiding places is to let your unconscious work on them while you sleep. Ask yourself one of the story prompts before you go to bed, sleep on it, and see what turns up in the morning. This reminds me of a joke my Glaswegian friend Fraser told me. What kind of cheese can get a bear to come out of its cave? *Camembert.* (Say it out loud and you'll get it.)

### 2. You can spot stories in the wild.

Start to train your story ears by listening to other people relating stories unselfconsciously. People do this all the time, and you'll pick up stories in the tea room, in the corridors, in meetings. When you hear someone say something like, 'Last week I was over in Marketing talking with Bronwyn when . . .', you know you're about to hear a story. Stories are a recounting of an event and start with a time, a place and a person. If you listen for them, you'll start to hear stories everywhere.

### 3. You can run a story circle.

This just means you get a small group of people together (usually sitting in a circle so everyone can see each other) to swap stories. When one person starts to tell a story it usually makes the other people in the group think of stories of their own to add, which means this environment draws out stories very naturally. The prompts in the next sections work well with story circles. If you'd like to know more about story circles, an excellent guide is *Circle of the 9 Muses* by David Hutchens.

### 4. You can interview people for stories.

You may need to do this with customers when following up on case studies to use as success stories. It's also an easy alternative if it's impractical to get people together for a story circle. Interviews can feel stilted; the more conversational you keep it, the more natural it will feel. It can be difficult to get people to relate specific experiences because they're used to speaking in abstract business-speak, just as you are. 'Can you think of a time when that happened?' or 'Can you give an example of that?' are useful questions to have up your sleeve. Remember to tell people what you'll be using the story for, and make sure you have their permission if you will be using their name or identifying characteristics.

# 5  Identity Stories

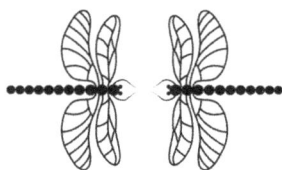

## Who we are

Identity stories are fundamentally about who you are, what's important to you and why you do what you do. They are your history, your values, your purpose.

People don't work just for money anymore. People work for meaning. To be engaged at work, they want to know their work serves a purpose, and they want to work with people who believe what they believe. Identity stories reinforce that sense of purpose far more than abstract mission or vision statements. They give people something concrete to relate these statements to.

All of the stories we'll look at in this book can apply at different levels. You can think of them as concentric circles, with you in the centre, your team in the next circle, and your organisation in the outer circle. Thinking of the identity stories in this section as they apply to you personally will help you with your first task—finding out why your work matters to you.

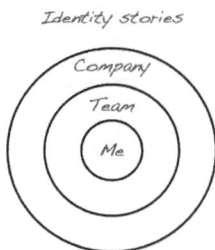

*Identity stories*

But these stories are also fantastic to explore on a team level, and doing this will help you with your third task—finding out why it matters to your team. I love working with teams as we draw out and look at these stories. They gain a much stronger sense of what their work is all about and a much greater appreciation of one another's role. It's always surprising what they discover.

Use identity stories to . . .

- engage your team and unite them behind a common purpose

- let your team and your customers know the 'why' behind what you do

- attract new staff members who are a great fit

- create a sense of cohesion and belonging in your team

- create a culture that embodies your most important values and takes them to the next level.

You'll know it's working if . . .

- your team are happier at work

- your team are more engaged at work

- your team engage in more of the behaviours that represent your core values or your transformation

- your team take greater responsibility for their actions

- your team innovate to take your values beyond expectations.

# History

I was born in the United States. Growing up, I was told many stories in school about the history of my country.

There was the story of the Pilgrims, a group of English people belonging to a persecuted religious sect who left everything behind to seek religious freedom in a new land. They landed at Massachusetts Bay and were nearly wiped out in their first year through illness, cold and starvation, but they were saved by befriending the native people, and they celebrated their survival at harvest with a three-day party that became the first Thanksgiving feast.

There was the story of the brave rebels of the original thirteen colonies, who threw off the yoke of English oppression and won a revolution against a much greater power through sheer determination and the leadership of the great general George Washington, and who collaboratively worked through their differences to build a new society based on democracy, a society of the people, by the people, for the people.

There was the story of the great Civil War between the North and the South, how brother fought against brother, how slavery was finally abolished and how President Abraham Lincoln, through conviction, empathy and honesty, brought about peace and reunited the nation.

Peppered through these larger stories are anecdotes about people who loom large in American mythology: Benjamin Franklin discovering electricity while flying a kite attached to a key in a thunderstorm; Thomas Jefferson playing his violin to soothe his soul as he wrote the Declaration of Independence; Samuel Adams leading the charge to dump a shipment of tea into Boston Harbor to protest unfair taxes; John Hancock writing his signature on the Declaration of Independence large enough for King George to read it from England; Harriet Tubman, after herself escaping slavery, going back again and again into the South to lead other slaves to freedom.

I want to assure you that I do know—of course I know—that these stories as they were told to me as a child are more mythology than historical fact, and that they've been seriously spin-doctored over the years. I also know they're one-sided rather than inclusive, and that there are many, many other stories that show dark and unflattering sides of America. (I went to school in the 1970s and 1980s. I hope the versions they are telling kids these days are at least more inclusive.) But the mythological character of these stories is partly what makes them so powerful. These stories course through my veins. For better or worse, they are part of who I am. When I think about what it means to be American, I think of these stories. If what I see around me is congruent with the stories, I feel proud. If it's not, I feel sad, nostalgic or even angry for what has been lost.

My partner, Dom, is French, from a little village in Normandy. He has the same experience from his upbringing, except that the stories that run through his veins, the ones that make him stand a little taller when he hears the Marseillaise, are stories of Charlemagne, William the Conqueror (Guillaume le Conquérant) and the French Revolution. And the Visigoths. For some reason no one else can fathom, Dom finds great appeal in the Visigoths. He and his friend Pierre can argue for hours about whether or not the Visigoths contributed anything significant to French culture. Personally I think it's just an excuse to sit around and drink Cognac.

Why am I telling you this?

Because just as countries have stories that help to transmit their culture from one generation to the next and give their people a common identity, so do teams and organisations, and it's worth doing a bit of digging to find some of them.

Having a common sense of history helps everyone to feel part of the same team. If these stories are shared with new employees, these individuals will feel part of the team much faster. For a leader, the stories from your own personal history also help to define who you are and what you believe.

An example from business is the story of how Johnson & Johnson reacted to the Tylenol crisis of 1982. Tylenol is a brand of paracetamol sold in the United States. In the autumn of 1982, someone (it's still an unsolved case) replaced Tylenol Extra-Strength capsules with cyanide-laced capsules, resealed the packages, and put them on the shelves of several pharmacies and food stores in the Chicago area. Seven people died from consuming the capsules.

News coverage about poisoned Tylenol began before Johnson & Johnson knew what had happened. James Burke, chairman of Johnson & Johnson, reacted by forming a strategy team, and he gave them this guidance. First, 'How do we protect the people?' and second, 'How do we save the product?'

The company's first actions were to alert consumers via the news media not to consume any kind of Tylenol product. They stopped producing and advertising Tylenol, and they withdrew all capsules from store shelves nationwide. Six months later they reintroduced the product with new tamper-resistant packaging.

The story has become a case study for how an organisation should communicate with the public during a crisis. It's also a key story that defines the history of Johnson & Johnson as a company.

If I were an employee at Johnson & Johnson, it would make me feel good to know that my company has a history of caring for people first, before the brand. I would want to uphold that tradition by bringing that value to my own daily work.

### Prompts for history stories

*Prompts for exploring your own history:*

- Where did you spend your childhood and what was life like when you were a child?

- What people and events from your childhood had an impact on who you are today?

- What was true then that is still true today?

- How did you start your career?

- What made you choose that kind of work?

- What people or events during your career have had an impact on how you approach your work now?

*Prompts for exploring your team's or organisation's history:*

- Who founded your organisation and why? What need were they responding to?

- What were the early years like for your team or organisation? What was challenging?

- What was true then that is still true today?

- Who are some of the key figures in your team's or organisation's history? What are some things they did to make a difference?

- Did they have to make any hard decisions or lead people through a difficult period?

- Are there anecdotes about a time they did something outrageous or surprising in a meeting?

- Are there anecdotes about times they showed special care or empathy towards an employee or customer?

## Values

Most organisations have some values written into their mission or vision statements. They might say they stand for excellence, innovation, integrity, inclusiveness or outstanding customer service, to name just a few. You probably have some similar ones. But without any context, these are fuzzy, abstract concepts that mean different things to different people, and if they feel empty, shadow stories can quite easily form around them.

How does someone on your team know what these values mean and how to translate them into actions? Does smiling and greeting the customer constitute outstanding customer service? Does filling a diversity quota constitute inclusiveness?

Your team need some examples to make these abstract words concrete—true stories about how these values have been enacted.

Here's how one company did this. Nordstrom department stores in the United States say they are about 'a relentless drive to exceed expectations' and delivering 'the best possible shopping experience'. So far, that doesn't sound any different from the values espoused by thousands of other companies.

However, Nordstrom make it clear what they mean by collecting true and surprising stories that demonstrate the level of customer service they are talking about.

There's the one about the man who returned a set of snow tires to the Nordstrom store in Fairbanks, Alaska. Nordstrom don't sell tires. The man purchased them from the store that used to occupy the same retail space before Nordstrom moved in. The Nordstrom employee refunded his money anyway.

Or the one about a member of the housekeeping staff at a Nordstrom store in Connecticut who found a customer's bags, along with her receipt and flight itinerary, in the car park. He drove to JFK International Airport in New York and had the airport page her to let her know he had brought her bags to her.

These details make all the difference between empty values statements and ones that mean what they say. If someone asks a Nordstrom employee what it means to exceed expectations, they can say, 'Well, let me tell you . . .'

The stories also give employees the freedom to use their own judgement in turning the values into actions. Rather than providing rules and guidelines, the stories inspire Nordstrom staff. Nordstrom build on this culture by making time regularly for employees to share their best customer service stories.

You can do the same thing with your team's or your organisation's values, transforming them from empty statements into carriers of meaning.

In addition, your personal values as a leader are a big part of exploring why your work matters to you. It's worth the time to identify what these values are and how you have seen them play out in your own life and in the lives of people around you.

### Prompts for values stories

To discover stories about your own or your team's values, you can go one of two ways.

1. If you already know what the values are meant to be (exceeding expectations or excellence or innovation, say), you can find stories about where you've seen these things happening. Finding the stories will make the values more real. *Values* ⟶ *Stories*

2. Or you can reverse-engineer your values by exploring some stories first and then deciding what values these stories represent. This is a more organic, bottom-up approach. And you'll know where the values came from. *Stories* ⟶ *Values*

*Prompt for the Values* ⟶ *Stories method*:

- Think of a time you have seen someone demonstrating <value>, whether here or somewhere else. What happened?

*Prompts for the Stories* ⟶ *Values method*:

- Think of a time you felt particularly proud of your work or your team's work. What happened?

- Think of a time you felt you or your team missed an opportunity or missed the mark. What happened?

- Think of a time someone demonstrated a characteristic that your team or organisation needs, whether it was here or somewhere else. What happened?

A great way to tap values stories in your team is to use them to build momentum for changing behaviour. Here's how.

First, make time for your team to share these stories with each other, and make sure you capture and collect them. The stories serve as inspiration and as platforms for more ideas.

Next, work with the team to brainstorm ways to turn these values into behaviours and actions. For example, if your value is customer service, you might ask:

- What are some ways we could help customers get what they need faster and more easily?

- What are some ways we could build a better relationship with customers?

- What are some ways we could exceed our customers' expectations?

This should generate some ideas you could put into action.

Then keep looking for examples where people are doing these things and keep sharing the stories about them with the team. Every customer interaction now becomes a story opportunity. Every team meeting becomes an opportunity to build momentum on new behaviours.

When you are looking for opportunities to recognise and reward team members, you'll also now have the stories to base this recognition on, and you'll have a set of stories to spread outside your team to replace some of those shadow stories with more positive ones.

## Purpose

Purpose stories go to the very heart of making it matter. They are why you do what you do.

People want to know they're contributing to something meaningful, something important. Something that stirs their hearts and their imaginations. Purpose stories describe what that is. Purpose stories have a lot in common with values stories: they are both about what you believe. Values stories are about what characteristics are important to you—integrity or the pursuit of excellence, for example. Purpose is what gets you out of bed in the morning, the reason you want to do what you do.

When Blake Mycoskie was travelling in Argentina in 2006, he was struck by the hardships faced by children growing up without shoes. In response, he created TOMS shoes, a for-profit business, not reliant on donations, that gives a pair of shoes to a child in need for every pair of shoes their customers purchase. Now TOMS shoes have extended this model to give not only footwear but also glasses, clean water and more when customers purchase products from them.

People who work for TOMS know they're not just selling footwear, eyewear and coffee. They're improving lives. And the stories about some of the ways they're improving lives are published on the TOMS website (www.toms.com/where-we-give).

Improving lives is TOMS' purpose. It's why they do what they do.

Your purpose doesn't have to be this dramatic. It just has to reflect what you are really about. It has to mean something to you, your customers and the people who make it happen.

My partner, Dom, is a pâtissier and has his own pâtisserie and café. Dom believes in the joy that small pleasures bring. That's why he has been making beautiful cakes and pastries at his shop for 28 years. His team know they're not just serving coffee and croissants. They're making people feel good, putting a little bit of joy into their world, and there's a wall of happy customer selfies to prove it. The staff know most of those customers because they're regulars, and when they look at those photos they know they've helped brighten these people's days a little. This gives them a greater sense of purpose than simply churning out 400 coffees every day.

### Prompts for purpose stories

*Here are some prompts to help you identify your purpose:*

- Think of a time you realised your work made a real difference to someone. What happened?

- Think of a time you noticed a genuine problem and realised you or your organisation *could* make a difference by doing something differently. What happened?

- Think of a time something made you feel angry or sad or helpless, and you had a strong desire to do something to change it. What happened?

- Think of a time something made you feel joyful or empowered or grateful, and you had a strong desire to do something to put more of that into the world. What happened?

Purpose stories often bring up strong emotions. This is what makes them so powerful; when people find their purpose it shakes up their world. Don't shy away from that, but you might want to warm up on some history or values stories first.

# 6  Insight Stories

## Who we serve

Insight stories are about understanding your customers. You can't make it matter if you don't know who it matters for. By knowing your customers better, you can develop the customer obsession that's needed to thrive in the Age of the Customer. You can learn how to create services and products they will love, that become an indispensable part of their lives.

Insight stories involve listening to real customers talk about their real experiences in order to learn what they do, how they do it, why it's important to them, what they risk, what's rewarding and what drives them crazy.

Use these stories to . . .

- bust assumptions you may have about your customers

- overcome barriers between your team and your customers

- stop wasting time trying to give your customers a solution they don't want

- give them something they'll truly value

- build trust

- create a culture of collaboration.

You'll know it's working if . . .

- your customers come to you more often for advice

- projects run more smoothly, with fewer misunderstandings

- your customers are happier with your services and with the outcomes of projects and solutions

- your team and your customers collaborate to create more innovative solutions.

## Giving your customers a gift

Let's say you want to give someone a gift. Maybe you're going over to their house for dinner and you want to bring them something. If you don't know them well, what do you give them? You might bring a bottle of wine or maybe some flowers. Choosing the wine can be tricky. Since you don't know them well, you might choose a decent bottle of red, one you like or one that's popular. There's a good chance they will like this. Even if they don't like it, they'll appreciate the fact that you thought to bring them a gift. But will it stand out in their memory? Will they feel that you selected it particularly for them? Not likely.

On the other hand, a few months ago I bought a gift for my friend Jan, whom I know well. It was a small orange resin dish from Dinosaur Designs. I knew she would like it because (a) I know she loves the bright warmth of oranges, pinks and reds, (b) I know that Dinosaur Designs is one of her favourite shops, and (c) I know she loves to entertain and set a stylish table.

It was fun to buy the gift for Jan and fun to anticipate giving it to her because I knew she would like it. When Jan opened the gift bag and unwrapped the tissue paper she said, 'It's from Dinosaur Designs! It's my favourite colour! I love it!' Her sincere reaction was worth any effort or expense I had gone to in purchasing it. It was so much better than giving her a generic bottle of wine.

Because I know Jan well, because I've spent time talking with her and paying attention to what she likes and doesn't like, I can give her something she loves without having to ask what she wants.

It's the same with your customers. If you take the time to get to know them well, you're able to create real value for them. You can still ask them what they'd like, but you'll also be able to anticipate their needs and not be limited to survey responses or requirements specifications. You might even be able to give them something they absolutely love that they didn't even know they needed.

This kind of engagement takes time and effort, but it's worth it.

## It's not like gathering requirements

When I work with groups on finding their customers' stories, there are two things that seem to be particularly difficult for them to get their heads around.

1. It's not like gathering requirements. You are not trying to find out specifically what people need from your service or solution. There's certainly a time for that, but that's not what this conversation is about. You are trying to find out about *them*—what they do, why it's important and why it's sometimes hard.

2. It's not about fishing for good case studies and PR stories. Again, there's a time for that, but this isn't it.

This is just about getting to know people better. It's about starting a conversation, opening a dialogue, building trust and sparking ideas.

Imagine Alec and Mary are out together on their first date. Alec likes Mary, but he's nervous and wants to give a good impression, so he starts talking a lot about his work. After a while he thinks maybe he shouldn't talk all the time and ought to ask Mary a question.

He looks into her eyes and asks, 'So Mary, what are you looking for in a relationship?'

Is this a good conversation opener from Alec?

No. This is icky. Alec has just ruined his chances. (If you had trouble with this question you may need to seek some dating advice.)

Alec isn't really asking about Mary with this question. He isn't trying to get to know her. He's just trying to get her to give him a list of boxes he can tick so he can say he meets the requirements. The question is more about him than it is about her.

This is what we tend to do when we survey or talk with our customers. Rather than being genuinely interested in them, we're much too quick to try to find out what they think about or need from us. Find out about them instead.

## Triumphs

Triumphs are stories that have had a happy ending for your customers. They are stories where your customers have succeeded in accomplishing their goal, possibly overcoming challenges along the way.

In these stories, you want to find out what success looks like for your customer. What is a specific example of a time they've achieved something satisfying? What happened? What were some things that made it difficult? What were some things that made it easier? Why was it important for them?

Note that you and your services may not come into this story at all. It might have nothing whatsoever to do with you. You are using it for context, to understand what they do, how they view their role, the kinds of things that are important to your customer and what makes them say, 'Yes!' It just helps you to understand them better. If you don't know what success means for your customer, how can you possibly help them to achieve it?

## Tragedies

Customer tragedies are the flip side of the success stories. They are times when your customer has missed an opportunity. They have been frustrated in trying to reach their goals, possibly because the challenges were too difficult to overcome. Again, your team or service may not come into this story at all. This is about your customer and their world. It's about understanding what frustrates them and gets under their skin.

## Trust

Now we've come to the point where there's a story about your customer interacting with you. This story might be good or bad, in the sense that the customer may have had a positive or a negative experience with you, or a mix of both. It's a trust story because it helps you gauge the level of trust the customer already has in your service.

People often have difficulties with just listening to this one because they are so used to wearing their problem-solving hat.

Remember, you're not trying to solve any problems at this point so don't raise their expectations about that, although you are likely to use the stories to help you work towards better solutions.

You're also not trying to extract a success story from them for marketing purposes. If you uncover a good case study from this process, go back to them later and ask if you can use it for that purpose. Right now you're just listening to understand what their experiences have been. You are just getting to know them. They'll appreciate it. You can even do it over a coffee.

### *Prompts for insight stories*

These story prompts are designed with internal customers in mind. You may need to make adjustments that make sense for your customers.

- Think of a time you felt particularly proud of your work or your team's work. What happened?

- Think of a time you felt you or your team missed an opportunity, perhaps because the challenges were too difficult to overcome. What happened?

- Think of a time you had an interaction with us (with our team or with our services). What happened?

- Think of a time you've seen or experienced something you think is needed here now. What happened?

- What drew you to the work you do? Was there something about it that particularly appealed to you, or a way you wanted to make a difference?

# 7 Influence Stories

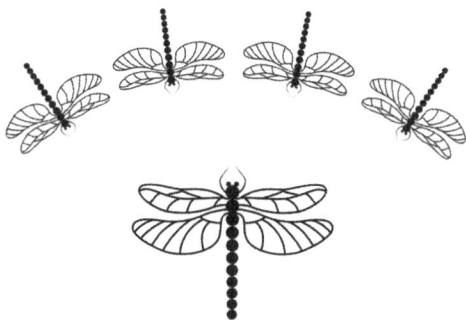

## Where we're going

Influence stories are fundamentally about change. They are about creating a compelling story that will inspire meaningful change. A story people want to be part of. We talked earlier about the story continuum—how some stories are more like plants growing in the wild and other stories are like garden plants that have been pruned to give them a pleasing shape. Influence stories need some thoughtful shaping.

Use influence stories to . . .

- create a change vision that inspires action

- address shadow stories about fear of and resistance to change

- help your team stay focused and committed through periods of change

- create a culture of innovation.

You'll know it's working if . . .

- when you present an idea for transformation, people want to engage with you in a conversation about how to shape it and take it forward

- your team wants to participate in the conversation about how to shape the transformation idea and take it forward

- the people who will make it happen act like it's their idea too

- people are willing to take action to make the change happen

- your team remains focused and engaged throughout the change period.

## The story of transformation

Change, whether it's digital transformation or any other kind of change, is hard. Change initiatives are notoriously fraught with problems. A number of studies have reported that around 70 per cent of change initiatives fail to achieve their objectives. Here is another interesting figure. On average, around 95 per cent of a company's employees either don't know what the strategy is or don't understand it.[4] How can people want to make a change happen if they don't even know what it is?

Successful change needs a number of things to make it work, but one of the first and most important is a vision, a story people tell themselves about the change, a story they want to be part of. The story needs to mean enough to them that they're willing to make changes—changes that may be difficult, changes that will push them out of their comfort zone, changes that may be downright frightening. People don't like to make these kinds of changes. They like to stay safe. A pie chart is probably not going to do the trick.

For this we have to turn to the kinds of stories that have inspired people to change for millennia. There is a story pattern that is so familiar to us and so irresistible that it's almost part of our DNA. It's in our blood, in our cells. It is the story of the human experience of change, of risks, of struggle and fortitude, of triumph or failure. It gives us a script to follow through periods of change and chaos.

It's known as the arch plot, or the quest, or the hero's journey. Joseph Campbell introduced the concept in 1949 in *The Hero with a Thousand Faces*. Here is a simplified version of how it goes. We'll use *The Lord of the Rings* as a classic example.

4 Robert S. Kaplan and David P. Norton. 'The Office of Strategy Management'. *Harvard Business Review*, October 2005.

The keys to this story are that there is a current reality (**what is**) and a desired future state (**what could be**). There is something very important at stake that means it's imperative to make **the journey** from the current reality to try to reach the desired future, even though this journey is difficult and filled with risks and peril. There's no real choice; it must be attempted.

In *The Lord of the Rings*, Frodo's current reality is that his uncle Bilbo has handed down to him a mysterious ring that turns out to be the One Ring of Power created, lost and sought by the evil sorcerer Sauron. Things may feel normal in the Shire now but in reality it's under threat by outside forces. If the Ring is destroyed, Sauron loses all of his power, the world of Middle-earth is safe from him, and the Hobbits can keep living peacefully in the Shire. That's the desired future state. What's at stake is that if Sauron gets his hands on the Ring before it can be destroyed, he'll become all-powerful and turn Middle-earth into a nightmarish dystopia where everyone is his slave. There's no way for Frodo to offload his responsibility. Even though he doesn't want to, even though he's scared out of his mind, he has to undertake the journey to try to destroy the Ring by dropping it into the fiery cracks of Mount Doom in faraway Mordor. When the Black Riders come looking for Frodo in the Shire, his mission suddenly becomes urgent, and he's forced out of his complacency and procrastination.

So the hero responds to the call to adventure and commits to the journey. Along the way there are many obstacles, trials and dangers, and there is at least one powerful enemy, demon or dragon standing in the way of reaching the goal. Often the most difficult enemy to fight is within the hero him- or herself. No matter how difficult it gets, the hero doesn't stop but struggles resolutely to reach the goal, usually with help from some friends.

On his journey to Mordor, Frodo and his friends face many

obstacles and are nearly killed many times, but Frodo keeps going. What is at stake is too important to allow him to give up.

Ultimately the hero faces a final ordeal and prevails. (Or dies, but that's not the version you need for your digital transformation.) The desired future state is achieved, the world is transformed, and there's a big party. The hero has also been transformed and is not the same person he or she was at the beginning.

Frodo faces his final ordeal on Mount Doom and, with help, succeeds in destroying the One Ring, and Sauron. Sorry for the spoiler. There is a huge celebration as Aragorn is crowned king. Frodo goes back to the Shire and kicks out the riffraff. The Shire is at peace once more, but Frodo has changed. He's not the same Hobbit he was and can't settle down to life in the Shire. He knows he must sail to the West with Gandalf, Bilbo and the Elves.

If you watch a good movie or read a good book that follows this pattern, you'll always find yourself cheering at the end. Hearing stories like these can convince us to take risks and make changes in our own lives, if what's at stake is important enough to us. Remembering how the hero had to struggle can help us to get through the difficult parts of the journey.

But can you apply this framework to the story of your digital transformation vision and strategy? Yes, you can.

## What is

The first part of the story to find to influence change should show the current reality and why it's not satisfactory.

In *Leading Change* and again in *The Heart of Change,* John Kotter argues that one of the most common reasons change initiatives fail is in not creating a strong enough sense of urgency. People quite simply don't appreciate the need for change. He says 'people change what they do less because they've been given *analysis* that shifts their *thinking* than because they've been shown a *truth* that influences their *feelings*.' It's no good giving people data and facts and expecting the change to happen that way. You'll need the data and analysis to support your reasons for change, but they won't help people make the decision to leap. You need to reach them where they can feel it.

People need first to have an appreciation of why change is needed, and this needs to hit them at an emotional level.

So the first story is the story of now, the story of what is.

Here are two examples from *The Heart of Change* showing how this sense of urgency was created.

In one case, the new head of finance wanted to transform procurement practices in the company. He knew they could save millions of dollars this way, and leverage much better arrangements with their suppliers. He hired a student intern to look for something that every department used, and to find out the details of how that type of product was currently being purchased.

She chose gloves. She brought together one example of each type of gloves the company was purchasing, and she labelled them with where they were purchased and for what price. The company was purchasing over 400 different types of gloves.

The head of finance called the executive team together and dropped the pile of 400 pairs of gloves onto the boardroom table. The visual representation struck home. Many of the gloves looked identical even though they had widely different prices.

The pile of gloves was so successful in getting people to appreciate the need for change that it went on tour around the company, and long afterwards people continued to tell each other the story of the gloves.

In another example from Kotter, a leader made a video recording of an angry customer describing the service he had received. Most people who didn't have direct contact with customers did not appreciate what the customer experience was like. Hearing this customer tell his story in his own words struck home and was the catalyst for needed change.

Remember Frodo, procrastinating about leaving the Shire until the Black Riders were hot on his heels? You need something to make the imperative for change real for people, otherwise there's no reason for them move out of their comfort zone.

## Prompts for what is stories

To get these stories, look for real examples in your organisation that will demonstrate, in a way no one can ignore, why change is needed.

Look for ways to have customers describe these in their own words, or ways to physically show people what is happening and why it can't continue.

*Prompts for what is stories:*

- Think of a situation or example in your organisation that demonstrates the need for transformation. What is happening?

- Think of a situation in an organisation similar to yours that demonstrates the need for your organisation to transform. What is happening?

- How can you make these stories real for people so they feel an urgent need to act?

## What could be

It's critical to create a sense of urgency with the what is story, but you can't leave people there. A negative story is useful as a wakeup call, but by itself it doesn't work to inspire positive action.

The next story, what could be, is the companion to what is. It's the good cop. What could be offers people a vision of how things could be different. This is the desired future state, the change you want to inspire.

For this story, you need to find an illustration where the change has already been implemented successfully. Or part of it, or something like it, has. It's important that it be a real example. Remember that what allows people to believe they can succeed are stories about other people who have done it before them.

Here's a real example of a what could be story that might be closer to the kinds of problems you are facing.

In 1996 an Australian lawyer named Steve Denning was working for the World Bank. Following changes in top management he was sidelined. From being Director of the Africa Region, he found himself with no real role. His boss suggested offhandedly that he might look into information.

The World Bank is a lending organisation. They try to address issues of poverty by lending money to projects in developing countries. They did not at that time see themselves as an information organisation. Being told to look into information was like being sent to Siberia.

However, because he wasn't ready to leave the World Bank, he started looking into information. And the more he looked, the more potential he started to see for how the World Bank could use its information more effectively to support its mission of alleviating poverty. They had a great deal of information that could be of use to people in solving their own problems, but no one had any access to it.

He discovered a model in how the U.S. Centers for Disease Control and Prevention had created a website to allow the public to access health information. Remember, 1996 was early days for the public use of the internet.

Eventually Denning was granted ten minutes to talk to the Change Committee of the World Bank, and as part of his presentation he told a very short story about a health worker in a remote town in Zambia who went to the website for the Centers for Disease Control and Prevention and found the answer to a question about malaria. He asked the committee to consider what kind of organisation the World Bank could become if they used their information to help people in developing countries find the solutions to their own problems.

When he told this story, the committee members immediately wanted to engage with him in a conversation about how they could make this happen at the World Bank. After weeks of knocking on doors and lobbying unsuccessfully about his idea, he suddenly found these people treating it as if it were their idea.

He resented their claiming his idea—until he realised this was actually exactly what he had been trying to achieve. It was the beginning of a long and difficult change process, but it provided the needed catalyst. Four years later, knowledge management was an official part of the World Bank—in its mission, its organisational chart and its funding. Denning had brought about a transformation of a huge, conservative organisation.

You can read Denning's full story and find the text of his presentation in his book, *The Secret Language of Leadership*.

The story he used in the presentation was extremely short—only a couple of sentences. Denning became a storytelling proponent and advocates keeping these stories as minimalist as possible in order to let the audience see themselves in the story and claim the idea as their own.

As part of this process, it's also important to invite the listener at the end of the story to imagine what would happen if that change were implemented here.

### Prompts for what could be stories

You need a real example where the change you are seeking to bring about (or something similar) has already (at least partly) happened successfully somewhere. This is so the people you are trying to bring on board can see themselves in the story and how it might apply to them.

When you are dealing with innovative changes, it can take some work and imagination to find the example you need. It might come from a different company, a different industry, even a different time. Keep looking until you find one that works.

Fortunately, a number of case studies of successful digital transformation programs are coming to light in Australia and New Zealand. These can serve as the basis for your story.

A very important aspect of this story is that it must have appeal for your particular audience. The desired future state (what could be) needs to be something that matters to them. They need to agree it's a goal worth striving for. It needs to relate to their purpose, why they do what they do. You need to be able to make the vision of this future state clear.

*Prompts for what could be stories:*

- Think of an example where something similar to this change has already occurred, at least partly. What happened?

- What will your organisation be like if you succeed in the transformation? What things will change for the better?

## The journey

In this part of the change story, you'll let people know what will be involved in getting from the current reality to the desired future. It will be a difficult journey with many challenges. Some of the obstacles will be external, such as time pressure and logistics, and some will be internal, such as people's own fears and doubts or the need to collaborate with people they haven't worked with before.

Try to identify all of these types of challenges, especially the internal ones, so you are able to address the unspoken shadow stories that may be forming. Acknowledge the reality of the challenges and let people know what support is available to help them on the journey. These are the people who will make the transformation happen, and they need to have an idea of what's in front of them.

Anyone who has been through a difficult change or has faced something frightening has a story to tell that can help others on this journey. For example, in 2013 I quit a secure, well-paid job in IT to strike out on my own as a storyteller and storytelling consultant. I know how frightening it is to scrap your safety net and do something you're not sure will succeed. I know what kind of practical and emotional hurdles are involved in starting up a business, creating and refining your ideas, and developing and offering new programs to clients. I know what a challenge it is to find the right combination of discipline, motivation, playfulness and self-care that will keep you going for the long haul. These are experiences I can offer to others on the path of something challenging. In turn, I'm inspired and supported by the stories and experiences of many wonderful people who have also made a journey like mine.

Find these stories of the journey and offer them as support to the people on your team. Often you'll be able to find stories from your own experience that will provide them with courage and confidence. That's what leaders do.

### *Prompts for journey stories*

*Here are few prompts for stories about the journey:*

- What are the practical challenges for this transformation? What will be the difficult aspects to it?

- Think of an example where someone was able to successfully overcome challenges like these. What happened?

- What might frighten people or make them uncomfortable about the change? What do they risk in putting it into action?

- Think of an example where someone (maybe you) faced a situation similar to this that was uncomfortable or frightening. What happened?

# 8  Impact Stories

## Where we've been

These stories are about looking back to make sense and meaning out of what you have achieved and what you have learned. Impact stories are your team's tribal artefacts of the transformation. They are your organisational memory. They are also fantastic marketing collateral, helping you to demonstrate value where quantitative measures may not show the whole picture.

Use impact stories to . . .

- show the impact and value your team provides

- spread positive perceptions of your team through the organisation

- improve performance

- expand the collective wisdom of the team

- create a culture of leadership.

You'll know it's working if . . .

- the organisational grapevine starts carrying more positive stories about your team and your work

- qualitative ratings of your performance and value increase

- your customers come to you as a trusted adviser and enabler

- your team's performance improves through learning from previous experiences

- your team's confidence increases

- your team demonstrate greater leadership in their work.

## Value, impact and learning

Influence stories looked forward. It's also important to look back throughout the journey to see where you've come from, over both short distances and long ones.

The stories of where you've been do a few things.

They let you make your successes visible, both within your team and to your customers and stakeholders. They're another way to show the value you are providing, and they increase the team's confidence.

They give you an opportunity to reflect on what didn't work so well, to make it easier to learn from it for the next time.

They are a source of inspiration for the team, especially when things didn't look like they were going to work out but then you tried something different and they did.

## Successes

You've started to implement your transformation. How do you show that it's working and you're making progress?

You probably have some quantitative measures you're tracking. These measures are great, but they don't show the whole picture.

Many things of value are difficult to measure quantitatively. In addition, the numbers, while they're useful (as a mathematician by education I'm a fan of data), don't move people's feelings. They don't show why it matters. Stories, on the other hand, are very good at moving feelings and very good at communicating examples that show real impact and value. They're not a replacement for measurement, but they're an essential complement to it.

Success stories serve a couple of purposes. First, they show you're making progress. This is important both for the decision-makers in your organisation and for your team.

Some projects can go for months without communicating any successes. When this happens, people start to form shadow stories about what's happening. Before you know it, a perception has built up that your group is failing to deliver.

By communicating stories of successes, you populate the organisational grapevine with the true, positive things that are happening. This helps to create positive perceptions among your stakeholders.

It is also highly motivating for your team. True success stories about the wins you are having are a great way to show your team they are making a real difference and to encourage them to keep going.

Second, success stories can help persuade other customers or people in your organisation that they want to be involved. Stories about how you've helped one customer help other potential customers to relate and see themselves in the story. They want to be part of it too.

## *Prompts for success stories*

These stories typically come from a combination of your team and your customers. Often someone on your team will identify the fact that there's a success story because they've been interacting with the customer, and they know from verbal feedback that the customer is happy. Following up with the customer will then provide you with the details you need to turn it into a success story.

Just sharing these stories informally within your team is a very powerful thing to do to move your digital transformation forward. There is nothing like the confidence and motivation boost that a little success can bring. Don't wait until the end of a project to do this. The key thing is to capture the stories as they happen, as frequently as possible.

If you are sharing success stories outside your team as case studies to demonstrate your impact and value, they do need to be shaped a little. The pattern is based on the same hero's journey pattern we looked at in chapter 7. The important thing to remember is that it is your customer, not you, who is the hero of these stories. The role you and your team play in the story is that of a helper or enabler. Your customer is the main character.

In the story, your customer has an important goal, but there are obstacles in the way of achieving it. This is a problem for them. It causes them pain, and there is something important at stake. Your team is helping them overcome the obstacles so they can achieve their goal, feel happy and make the world a better place.

That's the basic pattern. It means that when you follow up with the customer, you need to look to answer these questions:

- Who is your customer? What will help your audience relate to them?

- What is their goal? Why is it important?

- What is or was standing in the way of them achieving their goal?

- What did they do to try to overcome these obstacles?

- How did your team become involved? How did they help?

- What has the outcome been for the customer? Did they achieve their goal?

- What has changed for the customer? What has changed for the world or for your organisation? What has changed for your team?

## Failures

Why capture failures? Surely you don't want people to know about the times when you didn't succeed.

Yes, you do. You may not want to shout them from the rooftops like your success stories, but you don't want to sweep them under the carpet either. Here are some reasons for acknowledging and sharing your failure stories.

1.  You and your team learn from them. By being open about failure, without recrimination, you can find out what really happened so you can do better next time. Over time, performance improves, and the failure stories become excellent planning tools.

2.  It creates a culture in which people are empowered to take appropriate risks. You can't accomplish anything really great without being willing to fail some of the time, and facing this takes some courage. Sharing failure stories makes this courage part of the culture.

3.  Sharing failures promotes trust. Being open about what went wrong requires vulnerability, and this creates a culture of trust, both within your team and with your customers.

I belong to a community of thought leaders, led by the marvellous Matt Church and Peter Cook. There are around 150 people in the group and we have our own social media platform where we ask each other questions, share our knowledge and celebrate each other's successes. Every Friday is Failure Friday, and the only things we are allowed to post to the group on Friday mornings are our failure stories. This isn't a way of beating ourselves up, and there's no obligation to post a failure. It's a way of providing encouragement to the rest of the group. You're saying, hey, we're all human. Here's how I stuffed up this week. It's also a generous way of sharing knowledge because it helps others in the group to avoid similar pitfalls.

As a leader, it's very useful to sprinkle your own failure stories into meetings and presentations to create trust and to show you're human.

### Prompts for failure stories

Failure stories come from your team. Ask for them at meetings and make sure they know it's okay to admit what went wrong. The point is to learn from it. It may help to be light-hearted about it by doing your own version of Failure Friday—maybe once a week, maybe once a month. Definitely share failure stories at the end of a project and at key project milestones.

*Here's a prompt you can use to help recall failure stories:*

- Think of a time (this week or this month or during this project) when you feel you missed an opportunity, or a time that things didn't go as planned. What happened? What can we learn from this?

## War stories

War stories are stories that looked like they were going to end in failure but instead became successes. They were difficult, painful experiences that in the end had a positive outcome. Perhaps everyone in the team pulled together in a big effort, or they came up with an innovative solution that saved the day.

War stories combine the best elements of success and failure stories, and they have more drama than either of them, which makes them very compelling.

In Shakespeare's *Henry V*, young King Henry has led his army to France, to take back territory he believes should rightfully be England's. The troops have fought many battles and are weary and tattered and longing for home. It is the pre-dawn before the battle of Agincourt. The troops have had a sleepless night. They are facing a French army that outnumbers them five to one. They have no chance. However, they have done something innovative. They have sharpened wooden spikes and driven them into the ground with the points facing outwards, hoping to stop the French cavalry. And they've placed their archers behind these spikes where they can shoot from a place of protection.

The day of the battle is a holiday in England—Saint Crispin's Day. In the play, Henry, wandering amongst his troops, overhears one of his commanders wishing they had a few of the men who were at home in England sleeping in for the public holiday. Henry calls out,

> What's he that wishes so?
> My cousin Westmoreland? No, my fair cousin:
> If we are marked to die, we are enough

To do our country loss; and if to live,
The fewer men, the greater share of honour.
God's will! I pray thee, wish not one man more.

[...]

This story shall the good man teach his son;
And Crispin Crispian shall ne'er go by,
From this day to the ending of the world,
But we in it shall be remembered;
We few, we happy few, we band of brothers;

[...]

And gentlemen in England now abed
Shall think themselves accursed they were not here,
And hold their manhoods cheap whiles any speaks
That fought with us upon Saint Crispin's Day.

After this speech, the troops have renewed courage. They fight the battle, and at the end of the day the English have lost relatively few men, and the French are beaten. The soldiers who fought for England that day have a story to tell to their families and friends about prevailing against seemingly impossible odds.

When things have looked bad but they have pulled through, your team have war stories too. If these stories go unnoticed and unremarked, it's a missed opportunity to give your team the pleasure and the motivation from hearing the story told and celebrated.

## *Prompts for war stories*

Any time you have finished a project or campaign, and you feel like you've been hit by a truck, but at the same time you're elated because you've come out the other side with a great outcome, you have a war story. You just need to retell it. It reinforces the learning and your team's feeling of success. Project debriefs with your team are a good opportunity to do this. Sharing your war story can turn another boring meeting into the big party that always comes at the end of the hero's journey. Remember the Ewok party at the end of *Return of the Jedi*?

Long ago, when people lived in tribes and hunted in groups to bring down big game, the tribe would hold a feast after the hunt. A highlight of the feast was when the hunters retold and acted out the story of the hunt. The hunters got to celebrate their success and show how brave, determined and clever they had been. Younger members of the group, listening to the story, got to learn more about hunting techniques, which would help them when their turn came to join the hunt.

It's an approach worth considering for your post-project meetings.

*Here's a prompt for war stories:*

• Was there a time we thought we weren't going to make it? What happened? How did we pull through?

# Postscript:
# An Invitation

We've come to the end of the book, but every ending is also a beginning, and I hope for you this is just the beginning of your journey—or just a milestone along the way—to bring more meaning into your work and to help others around you find more meaning in theirs. A journey of making it matter.

The way we have been framing our work isn't working. Too many people are unmotivated and disengaged, every weekday morning dragging themselves to jobs that don't inspire them and don't bring out the best they have to offer. There is too much at stake to let this waste continue. Too much of our short lives wasted, too much talent not being used. Too many wicked problems in the world to solve, too much pain to ease, too much of the planet at risk, too much beauty waiting to be created and appreciated.

You have the opportunity to make a change, by exploring why your work matters to you, why it matters to your customers and why it matters to your team, and then turning ideas into actions. The bad news is you'll need to let go of some old habits, such as communicating only in analytical or abstract language, and you'll need to open up and expose more of what's truly important to you when you are at the office. This can be uncomfortable and frightening, I know. The good news is that as a human being, making meaning is your birthright, and you need only reconnect with the built-in language that does it automatically: the language of story.

Imagine what the world could be like, imagine what your organisation could be like, if everyone contributed wholeheartedly. It wouldn't be just about meeting KPIs and creating value for shareholders. You could take on challenges that matter. You could explore mysteries, make art, help people live healthier, happier, more connected lives, give the world more laughter and less violence, empower the helpless, help to preserve this planet and the species that call it home. There are endless possibilities. You could feel satisfied at the end of the day that you had done good work.

And you thought it was just IT.

## Further reading

If you're interested in exploring some of these topics further, here are a few books you might like.

Brown, Brené. *Daring Greatly: How the Courage to be Vulnerable Transforms the Way We Live, Love, Parent, and Lead.*
Penguin, 2013.
Opening up is tough. Here's some help.

Campbell, Joseph. *The Hero with a Thousand Faces.*
Princeton University Press, 1949.
It's the seminal work on the hero's journey, but
it is kind of hard going. You might want
to look for a gentler introduction.

Denning, Stephen. *The Secret Language of Leadership: How Leaders Inspire Action Through Narrative.*
Wiley, 2007.
How an Australian lawyer used storytelling
to transform the World Bank, and what he
learned about leadership in the process.

Duarte, Nancy. *Resonate: Present Visual Stories that Transform Audiences.* Wiley, 2010.
Excellent advice on how to use stories and story
structure for effective presentations.

Gottschall, Jonathan. *The Storytelling Animal: How Stories Make Us Human.*
New York: Mariner, 2013.
The evolutionary psychology of story.
A very interesting and enjoyable read.

Hutchens, David. *Circle of the 9 Muses: A Storytelling Field Guide for Innovators and Meaning Makers.*
Wiley, 2015.
An excellent resource for storytelling tools and processes.

Kahneman, Daniel. *Thinking, Fast and Slow.*
Penguin, 2011.
The role of intuition versus analysis in decision-making. Illuminating.

Kotter, John P. and Dan Cohen. *The Heart of Change: Real-Life Stories of How People Change Their Organizations.*
Boston: HBR Press, 2002.
Why people succeed and fail at change, the eight-step path to success, and real case studies for each step in the path.

Sinek, Simon. *Start With Why: How Great Leaders Inspire Everyone to Take Action.*
Penguin, 2009.
People don't buy what you do. They buy why you do it.

Whyte, David. *Crossing the Unknown Sea: Work as a Pilgrimage of Identity.*
New York: Riverhead, 2001.
If you're not familiar with David Whyte's work, it is my very great pleasure to introduce you. This is one of those books I read again every couple of years.

Wilson, Timothy D. *Strangers to Ourselves: Discovering the Adaptive Unconscious.* Harvard, 2002.
All the stuff that goes on in our minds below the tip of the iceberg.

## Acknowledgements

Some thanks are due to people who have helped me bring this book into being.

To my mentors Matt Church, Pete Cook, Mark Hodgson and the whole community at Thought Leaders Global, thank you for the 'bum glue' that helped me get it written and for inspiring me with your own work.

To the giants on whose shoulders I stand (just a few of whom are listed in the notes and reading list), thank you for giving me a better view.

To my editor Jem Bates and editor/layout designer/proof reader Lu Sexton, thank you for your words of encouragement and for helping me to make this a better book.

To Dom, thank you for being there.

## About the author

Patricia McMillan lives at the edge where two worlds meet: the world of innovation, technology and digital transformation; and the ancient world of storytelling, which speaks to our human need to connect with each other and create sense and meaning in our lives.

She's on an epic quest to liberate the world from soul-sapping work.

Patricia helps business leaders use the power of stories to engage their teams, connect with their customers, and transform their organisations.

Her background in mathematics and 20 years experience in information technology make her ideally placed to bring these skills to CIOs and information professionals.

Patricia is an entertaining speaker and a performing storyteller, accredited by the Australian Storytelling Guild NSW.

Learn more at www.patriciamcmillan.com.